Maths Problem Solving

Year 5

Catherine Yemm

Brilliant Publications

Other books in the same series:

Maths Problem Solving – Year 1   ISBN 10... 1 903853 745
                                  ISBN 13... 978-1-903853-74-0
Maths Problem Solving – Year 2   ISBN 10... 1 903853 753
                                  ISBN 13... 978-1-90385375-7
Maths Problem Solving – Year 3   ISBN 10... 1 903853 761
                                  ISBN 13... 978-1-903853-76-4
Maths Problem Solving – Year 4   ISBN 10... 1 903853 77X
                                  ISBN 13... 978-1-903853-77-1
Maths Problem Solving – Year 6   ISBN 10... 1 903853 796
                                  ISBN 13... 978-1-903853-79-5

Published by Brilliant Publications
1 Church View, Sparrow Hall Farm, Edlesborough,
Dunstable, Bedfordshire LU6 2ES

Sales and stock enquiries   Tel: 0845 1309200 / 01202 712910
                             Fax: 0845 1309300
Sales and payments   e-mail: brilliant@bebc.co.uk
                             website: www.brilliantpublications.co.uk
General enquiries: Tel: 01525 229720

The name Brilliant Publications and its logo are registered trade marks.

Written by Catherine Yemm
Cover and illustrations by Frank Endersby

ISBN 10... 1 903853 788
ISBN 13... 978-1-903853-78-8
© Catherine Yemm
First published in 2005
Printed in the UK by Lightning Source
10 9 8 7 6 5 4 3 2 1

Contents

Introduction

Maths Problem Solving Year 5 is the fifth book in a series of six resource books for teachers to use during the Numeracy lesson. It specifically covers the objectives from the Numeracy framework that are collated under the heading 'Solving problems'. Each book is specific for a particular year group and contains clear photocopiable resources, which can be photocopied onto acetate sheets to be viewed by the whole class or photocopied onto paper to be used by individuals.

Problem solving plays a very important part in the Numeracy curriculum and one of the reasons Numeracy is such an important subject is because the skills the children learn enable them to solve problems in other aspects of their lives. It is not enough to be able to count, recognize numbers and calculate, children need to be able to use problem solving skills alongside mathematical knowledge to help them succeed in a variety of 'real life' situations. Many of the problem solving skills and strategies that are needed do not come naturally so they have to be taught.

Problem solving is not an area which should be taught exclusively on its own but one which should be taught alongside other mathematical areas such as numbers and shape, space and measures. Children will benefit from being given opportunities to solve problems in other areas of the curriculum and away from the classroom as well as in their Numeracy lessons.

When teaching children how to solve problems, the Numeracy strategy refers to a number of points that need to be considered:

- The length of the problems should be varied depending on the age group. Children will benefit from being given short, medium-length and more extended problems.

- The problems on one page or in one lesson should be mixed so that the children do not just assume they are all 'multiplication' problems, for example, and simply multiply the numbers they see to find each answer.

- The problems need to be varied in their complexity: for example, there should be some one-step and some two-step problems, and the vocabulary used in each problem should differ.

- Depending on the age of the children the problems can be given orally or in writing. When given written problems to solve, some children may need help to read the words, although this does not necessarily mean that they will need help to find the answer to the question.

- The context of the problem should be meaningful and relevant to the children. It should attempt to motivate them into finding the answer and be significant to the time. For example, euros should be included as well as pounds.

This Year 5 resource is organized into three chapters: 'Making decisions', 'Reasoning about numbers or shapes' and 'Problems involving "real life", money or measures'. Each chapter contains six lessons, one to be used for each half term.

Making decisions

The objective outlined under the 'Making decisions' heading of the National Numeracy Strategy for Year 5 children is: Choose and use appropriate number operations to solve problems, and appropriate ways of calculating: mental, mental with jottings, written methods, calculator.

In this chapter the emphasis is on choosing and then using the correct operation to solve a given problem. In Year 5 the children are developing their adding, subtracting, multiplying and dividing skills and they need to understand that different problems will need different methods to solve them. They should be encouraged to make and justify decisions by choosing the appropriate operations to solve word problems, deciding whether calculations can be done mentally or with pencil and paper and be able to explain and record how the problem was solved. The children should be provided with an opportunity to tackle mixed problems so that they learn to think openly and make a decision depending on the vocabulary used and the question itself. If children are not taught these decisive skills then it is common for them to assume that, to find the answer to a question with two numbers, you just add or multiply the numbers. The questions set out in this chapter are mixed and the children could be required to use any of the four operations. The questions the children will answer are designed to enable them to practise solving problems in a variety of relevant contexts.

When the children are answering the questions encourage them to use mental strategies, make notes and use more formal written methods, and give them an opportunity to use a calculator.

This aspect of problem solving is closely correlated to objectives 72–73, 'Checking results of calculations'. After choosing and using the correct operation the children should be encouraged to use an appropriate method to check their results: checking with the inverse operation, adding in the reverse order, checking with an equivalent calculation, checking by approximating, or using knowledge of sums and differences of odd or even numbers.

Reasoning about numbers or shapes

The objectives outlined under the 'Reasoning about numbers or shapes', heading of the National Numeracy Strategy for Year 5 children are as follows:

■ Explain methods and reasoning orally and, where appropriate, in writing.
■ Solve mathematical problems or puzzles; recognize and explain patterns and relationships, generalize and predict. Suggest extensions by asking 'What if...?'

- Make and investigate a general statement about familiar numbers or shapes by finding examples that satisfy it. Explain a generalized relationship (formula) in words.

The activities that follow are a mixture of problems, puzzles and statements. Lessons 1, 3 and 5 are related to shape while lessons 2, 4 and 6 are related to number. When given a statement such as 'A multiple of 8 is both a multiple of 4 and a multiple of 2, the children should be encouraged to provide examples to prove the statement, for exampe, 2 x 8 = 16, 4 x 4 = 16, 8 x 2 = 16. Others will be obvious questions that just need an answer. The teacher should try to spend time talking to the pupils while they are working to allow them to explain their methods and reasoning orally and to provide an opportunity to ask questions such as, 'What if…?' The plenary session at the end of the lesson also provides an opportunity to do this.

Problems involving 'real life', money or measures

The objectives outlined under the 'Problems involving "real life", money or measures' heading of the National Numeracy Strategy for Year 5 children are:

- Use all four operations to solve simple word problems involving numbers and quantities based on 'real life', money or measures (including time), using one or more steps including making simple conversions of pounds to foreign currency and finding simple percentages.
- Explain methods and reasoning.

The activities in this section are typically 'word problems'. The contexts are designed to be realistic and relevant for children of a Year 5 age. The questions involve the operations of adding, subtracting, multiplication and division, and deal with money and measurements including time and everyday situations. The questions may be one-step problems or two-step problems.

The teacher should try to spend time talking to the pupils while they are working to allow them to explain their methods and reasoning orally. The plenary session at the end of the lesson also provides an opportunity to do this.

This aspect of problem solving is closely correlated to objectives 72–73, 'Checking results of calculations'. After choosing and using the correct operation the children should be encouraged to use an appropriate method to check their results: checking with the inverse operation, adding in the reverse order, checking with an equivalent calculation, checking by approximating, or using knowledge of sums and differences of odd or even numbers.

The lesson

Mental starter

In line with the Numeracy strategy the teacher should start the lesson with a 5–10 minute mental starter. This can be practice of a specific mental skill from the list specified for that particular half term or ideally an objective linked to the problems the children will be solving in the main part of the lesson. For example, if the problems require the children to add and subtract then it would be useful to spend the first 10 minutes of the lesson reinforcing addition and subtraction bonds or the vocabulary involved.

The main teaching activity and pupil activity

This book aims to provide all the worksheets that the teacher will need to cover this part of the lesson successfully. The whole class activity page provides examples of the types of problems that need to be solved. They are designed to be photocopied onto acetate to show the whole class using an overhead projector. The teacher will use the 'photocopiable answer sheet' to work through the examples with the class before introducing the class to questions they can try by themselves. The teacher should demonstrate solving the problem using skills that are relevant to the children's abilities, these could be drawings, counters, or number lines.

Once the children have seen a number of examples they will be ready to try some problem solving questions for themselves. Within each lesson there is a choice of three differentiated worksheets. The questions on these worksheets are basically the same, just the level of mathematical complexity varies. This ensures that the questions are differentiated according to the mathematical ability of the child. It will also ensure that when going through examples during the plenary session all children can be involved at the same time. For example, for a question involving the addition of three numbers, the children may have had to add three different numbers, but when the teacher talks through the question the fact that to solve the problem the children need to add will be the important point being reinforced. If the teacher feels that some pupils would benefit from having easier or more difficult questions then they could change the numbers on the worksheets to something more suitable.

The plenary

One of the important parts of solving problems is discussing with the children about how and what they did to solve the problems. After the children have completed the problems, the plenary can be used to:

- discuss the vocabulary used in the question
- discuss how the problem can be approached
- break down a problem into smaller steps
- list the operations or calculations used to solve the problem
- discuss whether the problem can be solved in more than one way
- discuss how the answers to the problems can be checked
- divulge the answers to a number of the questions.

Extension

Any children who complete their task relatively easily may need to be extended further. As well as being given the more challenging questions they could be asked to make up a question of their own, which should involve the same operations.

Resources

For some questions it will be useful to make a number of resources available to the children such as:

- Number lines to 100
- A selection of 2D and 3D shapes
- All coins
- Analogue clocks with moveable hands, digital clocks.

Answers

We have supplied answers to questions where possible, but there are some questions which have multiple answers or require class discussion. Some questions require the children to show their understanding by making up a story involving the figures mentioned, and some others are statements which require the children to give an example which supports the fact.

Photocopiable answer sheet

Photocopy onto acetate sheet and project on wall or screen.

I will need to _____

To help me I will use _____

My answer is _____

I will need to _____

To help me I will use _____

My answer is _____

I will need to _____

To help me I will use _____

My answer is _____

This page may be photocopied by the purchasing institution only.

© Catherine Yemm

www.brilliantpublications.co.uk

Maths Problem Solving – Year 5

9

Making decisions

Whole class activity

Make up a number story to reflect this calculation:

624 + 228 = 852

What operation does the * represent?
584 * 249 = 335

How can you check?

The head teacher has bought 268 new footballs for the school. There are 12 classes in the school. How many footballs can each class have? Will there be any left over?

© Catherine Yemm

Lesson
1a

1. What is the difference between 58 and 192?

• •

2. Joe's mum uses 8 apples to make 1 litre of apple juice. How many litres can she make from 24 apples?

• •

3. The train starts its journey with 152 passengers. At the first stop 37 get off. At the second stop double that number get off. How many passengers are still on the train?

• •

4. School starts at 9.05 am. How late will Ravi be if he doesn't arrive until 10.25 am?

• •

5. Natalie and Ryan are playing cards. Natalie scores 145 then 84, and Ryan scores 163 and 77. What are their total scores each?

• •

6. The Thomas family are going on holiday. It is 83 miles from their house to their hotel. They are three quarters of the way there. How many miles have they travelled?

• •

7. Felicity's paddling pool holds 18 litres of water. Her bucket holds 1.5 litres. How many times will she need to fill her bucket up to fill the pool?

• •

8. Ella started reading her library book on Monday. She read 32 pages. She read 27 more on Tuesday. How many pages does she have left to read if there are 75 pages in the book?

1. What is the difference between 158 and 392?

   •••

2. Joe's mum uses 8 apples to make 1 litre of apple juice. How many litres can she make from 32 apples?

   •••

3. The train starts its journey with 182 passengers. At the first stop 37 get off. At the second stop double that number get off. How many passengers are still on the train?

   •••

4. School starts at 9.05 am. How late will Ravi be if he doesn't arrive until 12.55 pm?

   •••

5. Natalie and Ryan are playing cards. Natalie scores 245 then 134, and Ryan scores 263 and 127. What are their total scores each?

   •••

6. The Thomas family are going on holiday. It is 135 miles from their house to their hotel. They are three quarters of the way there. How many miles have they travelled?

   •••

7. Felicity's paddling pool holds 24 litres of water. Her bucket holds 1.5 litres. How many times will she need to fill her bucket up to fill the pool?

   •••

8. Ella started reading her library book on Monday. She read 62 pages. She read 47 more on Tuesday. How many pages does she have left to read if there are 175 pages in the book?

1. What is the difference between 358 and 892?

• •

2. Joe's mum uses 8 apples to make 1 litre of apple juice. How many litres can she make from 72 apples?

• •

3. The train starts its journey with 282 passengers. At the first stop 87 get off. At the second stop double that number get off. How many passengers are still on the train?

• •

4. School starts at 9.05 am. How late will Ravi be if he doesn't arrive until 2.51 pm?

• •

5. Natalie and Ryan are playing cards. Natalie scores 245 then 334, and Ryan scores 263 and 427. What are their total scores each?

• •

6. The Thomas family are going on holiday. It is 198 miles from their house to their hotel. They are three quarters of the way there. How many miles have they travelled?

• •

7. Felicity's paddling pool holds 35 litres of water. Her bucket holds 2.5 litres. How many times will she need to fill her bucket up to fill the pool?

• •

8. Ella started reading her library book on Monday. She read 132 pages. She read 127 more on Tuesday. How many pages does she have left to read if there are 475 pages in the book?

Making decisions

Whole class activity

Mrs Tandy won £200 in a crossword competition. She decided to give 5% of her winnings to each of her 5 grandchildren. How much did she have left for herself?

Neela went to the cinema to see a film. The film finished at 21:37. It was 92 minutes long. At what time did the film start?

The tanker brings 459 litres of petrol to the garage on a Monday morning. By lunch time two thirds of it have been used up. How much is left?

1. In the book corner of class 5 there are 9 book shelves. 3 of the shelves each have 15 books on them, another 3 each have 21 books on them and the last 3 shelves have 8 books on them each. How many books are in the book corner?

• •

2. If I increase 182 by 25, what is the new number?

• •

3. There are 12 pairs of scissors in a box. If the school secretary orders 7 boxes, how many new scissors will arrive at the school?

• •

4. Make up a number story to reflect the calculation:

$$39 - 18 = 21$$

• •

5. Joseph buys a bag of carrots weighing 8kg and a bag of potatoes weighing 3 times as much. How much do his vegetables weigh altogether?

• •

6. Peter and Jill are making a kite. Peter has a piece of string 185cm long and he cuts off 1m and 15cm to make a tail. How long will Jill's kite tail be if she uses the string that is left?

• •

7. What operation does the * represent?

$$13 * 11 = 143$$

How can you check?

• •

8. The manager of the local supermarket has promised to split any profit the shop makes between the workers. In the last year the shop has made £420 profit. If there are 35 workers at the supermarket, how much will they have each?

Lesson 2b

1. In the book corner of class 5 there are 9 book shelves. 3 of the shelves each have 25 books on them, another 3 each have 31 books on them and the last 3 shelves have 18 books on them each. How many books are in the book corner?

· ·

2. If I increase 182 by 56, what is the new number?

· ·

3. There are 12 pairs of scissors in a box. If the school secretary orders 11 boxes, how many new scissors will arrive at the school?

· ·

4. Make up a number story to reflect the calculation:

$634 - 89 = 545$

· ·

5. Joseph buys a bag of carrots weighing 16kg and a bag of potatoes weighing 3 times as much. How much do his vegetables weigh altogether?

· ·

6. Peter and Jill are making a kite. Peter has a piece of string 240cm long and he cuts off 1m and 15cm to make a tail. How long will Jill's kite tail be if she uses the string that is left?

· ·

7. What operation does the * represent?

$23 * 11 = 253$

How can you check?

· ·

8. The manager of the local supermarket has promised to split any profit the shop makes between the workers. In the last year the shop has made £1085 profit. If there are 35 workers at the supermarket, how much will they have each?

© Catherine Yemm

1. In the book corner of class 5 there are 9 book shelves. 3 of the shelves each have 7 books on them, another 3 each have 51 books on them and the last 3 shelves have 28 books on them each. How many books are in the book corner?

• •

2. If I increase 182 by 86, what is the new number?

• •

3. There are 12 pairs of scissors in a box. If the school secretary orders 19 boxes, how many new scissors will arrive at the school?

• •

4. Make up a number story to reflect the calculation:

634 – 289 = 345

• •

5. Joseph buys a bag of carrots weighing 28kg and a bag of potatoes weighing 3 times as much. How much do his vegetables weigh altogether?

• •

6. Peter and Jill are making a kite. Peter has a piece of string 540cm long and he cuts off 3m and 15cm to make a tail. How long will Jill's kite tail be if she uses the string that is left?

• •

7. What operation does the * represent?

43 * 11 = 473

How can you check?

• •

8. The manager of the local supermarket has promised to split any profit the shop makes between the workers. In the last year the shop has made £1820 profit. If there are 35 workers at the supermarket, how much will they have each?

Making decisions

Whole class activity

Mrs Potts has bought some chocolate biscuits for the school staffroom. She buys a packet of 50 and by lunch time on Monday 20% of them have been eaten. How many are left?

How many more than 892 is 1051?

Joanne is buying some apples for the school tuck shop. She could buy 30 separate apples at 32p each, or she could buy 30 in a bag for £8.40. Which would be cheaper and by how much?

© Catherine Yemm

1. Laura goes swimming on Monday and swims for 1 hour and 10 minutes. If she gets into the pool at 5.30 pm, what time does she get out?

• •

2. What operation does the * represent?

256 * 4 = 64

How can you check?

• •

3. The school car park covers an area of 120m². If 20 cars can park in the car park, what area does each space cover?

• •

4. The local pet shop has 8 rabbits, 4 gerbils and 1 lizard. How many legs do they have between them?

• •

5. The lorry which delivers the school stationery usually brings 84kg of boxes each week. This time a box weighing 38.5kg fell off the back of the lorry as it was travelling. How heavy was the load that arrived at the school?

• •

6. Lucy wants to buy a bike but she has to save her pocket money. She can reserve the bike if she puts down a 50% deposit. If the bike is £84, how much deposit will she have to put down?

• •

7. Make up a number story to reflect the calculation:

32.8 x 11 = 360.8

• •

8. Kasi is helping her mum clear up after her birthday party. There were 20 people at her party and 3 left 40ml of squash in their cups and 2 left 30ml of squash in their cups. How much squash had to be thrown away?

1. Laura goes swimming on Monday and swims for 2 hours and 25 minutes. If she gets into the pool at 5.30 pm, what time does she get out?

. . . .

2. What operation does the * represent?

   448 * 7 = 64

   How can you check?

. . . .

3. The school car park covers an area of 240m². If 20 cars can park in the car park, what area does each space cover?

. . . .

4. The local pet shop has 14 rabbits, 8 gerbils and 2 lizards. How many legs do they have between them?

. . . .

5. The lorry which delivers the school stationery usually brings 125kg of boxes each week. This time a box weighing 38.5kg fell off the back of the lorry as it was travelling. How heavy was the load that arrived at the school?

. . . .

6. Lucy wants to buy a bike but she has to save her pocket money. She can reserve the bike if she puts down a 50% deposit. If the bike is £124, how much deposit will she have to put down?

. . . .

7. Make up a number story to reflect the calculation:

   32.8 x 22 = 721.6

. . . .

8. Kasi is helping her mum clear up after her birthday party. There were 20 people at her party and 5 left 40ml of squash in their cups and 4 left 30ml of squash in their cups. How much squash had to be thrown away?

www.brilliantpublications.co.uk
Maths Problem Solving – Year 5

1. Laura goes swimming on Monday and swims for 3 hour and 25 minutes. If she gets into the pool at 4.40 pm, what time does she get out?

• •

2. What operation does the * represent?

1088 * 17 = 64

How can you check?
• •

3. The school car park covers an area of 360m². If 18 cars can park in the car park, what area does each space cover?

• •

4. The local pet shop has 21 rabbits, 12 gerbils and 8 lizards. How many legs do they have between them?

• •

5. The lorry which delivers the school stationery usually brings 225kg of boxes each week. This time a box weighing 78.5kg fell off the back of the lorry as it was travelling. How heavy was the load that arrived at the school?

• •

6. Lucy wants to buy a bike but she has to save her pocket money. She can reserve the bike if she puts down a 50% deposit. If the bike is £324, how much deposit will she have to put down?

• •

7. Make up a number story to reflect the calculation:

62.8 x 22 = 1381.6

• •

8. Kasi is helping her mum clear up after her birthday party. There were 20 people at her party and 9 left 40ml of squash in their cups and 6 left 30ml of squash in their cups. How much squash had to be thrown away?

Making decisions

The cost for one adult to go to the zoo is £2.50. It is half price for a child to go. How much would it cost for 4 adults and 4 children to visit the zoo?

The local swimming pool has a balcony each side for spectators to watch. The balcony on the right holds 176 people and the one on the left holds 83 people. How many people can watch the swimmers from the balconies?

All of the children in class 5 weighed themselves and the total was 625kg. If there are 25 children in the class, what is the average weight of each child?

1. Holly wants to go for a ride in a hot air balloon but she has to be 12 years old. She is 11 at the moment and her birthday is on 17th September. If it is the 22nd August today, how long does she have to wait?

2. Class 5 are making cards with beads. Mr Smith asks James and Jivin to sort the beads into colours. How many beads are in each pile if they start off with 150 and sort them into equal piles of yellow, red, orange, blue, green and white?

3. What operation does the * represent?

   108 * 169 = 277

   How can you check?

4. If the school tuck shop sells 20 apples at 10p each and 5 bananas at 22p each, how much money does it make?

5. Make up a number statement to reflect the calculation:

   340 ÷ 8 = 42.5

6. A recipe to make carrot cake uses 3.5 teaspoons of oil. If we want to make 4 carrot cakes, how much oil will we need?

7. How many altogether are 82, 66 and 103?

8. It is 175 days until Megan goes on holiday. It is 132 days until Megan's birthday. How long after her birthday will she go on holiday?

1. Holly wants to go for a ride in a hot air balloon but she has to be 12 years old. She is 11 at the moment and her birthday is on 17th September. If it is 22nd July today, how long does she have to wait?

2. Class 5 are making cards with beads. Mr Smith asks James and Jivin to sort the beads into colours. How many beads are in each pile if they start off with 270 and sort them into equal piles of yellow, red, orange, blue, green and white?

3. What operation does the * represent?

408 * 169 = 577

How can you check?

4. If the school tuck shop sells 30 apples at 10p each and 15 bananas at 22p each, how much money does it make?

5. Make up a number statement to reflect the calculation:

548 ÷ 8 = 68.5

6. A recipe to make carrot cake uses 6.25 teaspoons of oil. If we want to make 4 carrot cakes, how much oil will we need?

7. How many altogether are 142, 66 and 203?

8. It is 217 days until Megan goes on holiday. It is 132 days until Megan's birthday. How long after her birthday will she go on holiday?

1. Holly wants to go for a ride in a hot air balloon but she has to be 12 years old. She is 11 at the moment and her birthday is on 17th September. If it is 22nd April today, how long does she have to wait?

• •

2. Class 5 are making cards with beads. Mr Smith asks James and Jivin to sort the beads into colours. How many beads are in each pile if they start off with 540 and sort them into equal piles of yellow, red, orange, blue, green and white?

• •

3. What operation does the * represent?

608 * 269 = 877

How can you check?

• •

4. If the school tuck shop sells 30 apples at 13p each and 15 bananas at 27p each, how much money does it make?

• •

5. Make up a number statement to reflect the calculation:

676 ÷ 8 = 84.5

• •

6. A recipe to make carrot cake uses 9.35 teaspoons of oil. If we want to make 4 carrot cakes how much oil would we need?

• •

7. How many altogether are 242, 166 and 303?

• •

8. It is 317 days until Megan goes on holiday. It is 132 days until Megan's birthday. How long after her birthday will she go on holiday?

This page may be photocopied by the purchasing institution only.

© Catherine Yemm

www.brilliantpublications.co.uk

Maths Problem Solving – Year 5   25

Making decisions

Tegan bought a new book for £6.20. She gave the shopkeeper a £10 note but he had to give her the change in 20p pieces. How many did he give her?

Ellen goes to bed at 6.50 pm on Friday evening and wakes up at 7.17 am on Saturday morning. How much sleep has she had?

Harry and Sanjeev are 2 months and 4 days different in age. If Harry's birthday is 3rd January and he is older, when is Sanjeev's birthday?

Can you work out Sanjeev's birthday if he were the eldest?

1. On Monday the milkman leaves 2 pints of milk at all the even numbered houses inbetween numbers 46 and 56, and 3 pints of milk at all the odd numbered houses inbetween 51 and 73. How much milk does he deliver that day?

2. Take 284 and add a number. The answer is 342. What is the number?

3. Susan is going on holiday. Her suitcase weighs 23.4kg. The limit to go on the plane is 12.2kg. How much over the limit is she?

4. What operation does the * represent?

   33.25 * 25.43 = 7.82

   How can you check?

5. At midday it is 16.4°C in town. At night the temperature drops by 21°C. What temperature is it now?

6. Rani has a snake for a pet. Her snake is 54cm long. It could grow up to 6 times as long. How long could it be when it is fully grown?

7. Make up a number statement to reflect the calculation:

   45.2 x 3 = 135.6

8. Sam buys a lollipop for 45p and two ice creams which cost £1.20 each. How much money does he pay?

This page may be photocopied by the purchasing institution only.

© Catherine Yemm

www.brilliantpublications.co.uk

Maths Problem Solving – Year 5   27

1. On Monday the milkman leaves 4 pints of milk at all the even numbered houses inbetween numbers 46 and 56, and 3 pints of milk at all the odd numbered houses inbetween 51 and 73. How much milk does he deliver that day?

2. Take 484 and add a number. The answer is 602. What is the number?

3. Susan is going on holiday. Her suitcase weighs 25.4kg. The limit to go on the plane is 15.0kg. How much over the limit is she?

4. What operation does the * represent?

   63.25 * 25.43 = 37.82

   How can you check?

5. At midday it is 26.4°C on the beach. At night the temperature drops by 31.2°C. What temperature is it now?

6. Rani has a snake for a pet. Her snake is 54cm long. It could grow up to 8 times as long. How long could it be when it is fully grown?

7. Make up a number statement to reflect the calculation:

   45.2 x 6 = 271.2

8. Sam buys 3 lollipops for 45p and 2 ice creams which cost £1.20 each. How much money does he pay?

1. On Monday the milkman leaves 4 pints of milk at all the even numbered houses inbetween numbers 46 and 56, and 6 pints of milk at all the odd numbered houses inbetween 51 and 73. How much milk does he deliver that day?

••

2. Take 484 and add a number. The answer is 802. What is the number?

••

3. Susan is going on holiday. Her suitcase weighs 35.4kg. The limit to go on the plane is 20.6kg. How much over the limit is she?

••

4. What operation does the * represent?

    93.25 * 25.43 = 67.82

    How can you check?

••

5. At midday it is 36.4°C on the beach. At night the temperature drops by 47.2°C. What temperature is it now?

••

6. Rani has a snake for a pet. Her snake is 54cm long. It could grow up to 12 times as long. How long could it be when it is fully grown?

••

7. Make up a number statement to reflect the calculation:

    45.2 x 8 = 361.6

••

8. Sam buys 6 lollipops for 45p and 7 ice creams which cost £1.20 each. How much money does he pay?

Making decisions

Jennifer has been practising her sprinting. At the moment she can run around the track in 3 minutes and 54 seconds. She hopes to knock 95 seconds off her time. What time is she aiming for?

Christopher has been given £58.75 for his birthday. He buys a new bag for £5.40 and a new pair of trainers for £23.80. How much birthday money does he have left?

The local post office usually sells 534 first class stamps in a week. At Christmas it doubles its sales. How many stamps might it sell in a week at Christmas time?

1. Ashon buys 2 boxes of sweets containing 120 sweets each and 2 boxes of sweets containing 75 sweets each. If the sweets cost 2p each, how much does it cost him?

• •

2. Make up a number statement to reflect the calculation:

   456 – 163 = 293

• •

3. The patio in Ben's garden is 1m and 45cm long. His dad is going to extend it by 208cm. How long will it be then?

• •

4. To go camping for the weekend it costs £13 for the tent, £2 for the car, £3.50 for each adult and £2.20 for each child. How much will it cost 2 families with 2 parents and 2 children each to camp for the weekend if they go in their separate cars but share a tent?

• •

5. What three numbers could have a total of 172?

• •

6. The local bakery always donates its leftover bread to three local cafés. On Saturday they had 65 bread rolls, 49 chelsea buns and 72 french sticks left. If the bakery gives each café an equal total of items, how many did each café receive?

• •

7. What operation does the * represent?

   25 * 6 = 150

   How can you check?

• •

8. The pond in Trudy's garden holds 135 litres of water. In the summer 67 litres evaporate in the hot weather. How much water is left in the pond?

Lesson
6b

1. Ashon buys 3 boxes of sweets containing 120 sweets each and 4 boxes of sweets containing 75 sweets each. If the sweets cost 3p each, how much does it cost him?

• •

2. Make up a number statement to reflect the calculation:

656 – 163 = 493

• •

3. The patio in Ben's garden is 3m and 45cm long. His dad is going to extend it by 208cm. How long will it be then?

• •

4. To go camping for the weekend it costs £23 for the tent, £3 for the car, £6.50 for each adult and £3.20 for each child. How much will it cost 2 families with 2 parents and 2 children each to camp for the weekend if they go in their separate cars but share a tent?

• •

5. What three numbers could have a total of 372?

• •

6. The local bakery always donates its leftover bread to three local cafés. On Saturday they had 128 bread rolls, 81 chelsea buns and 67 french sticks left. If the bakery gives each café an equal total of items, how many did each café receive?

• •

7. What operation does the * represent?

25 * 9 = 225

How can you check?

• •

8. The pond in Trudy's garden holds 267 litres of water. In the summer 131 litres evaporate in the hot weather. How much water is left in the pond?

1. Ashon buys 4 boxes of sweets containing 120 sweets each and 5 boxes of sweets containing 75 sweets each. If the sweets cost 4p each, how much does it cost him?

• •

2. Make up a number statement to reflect the calculation:

856 – 263 = 593

• •

3. The patio in Ben's garden is 3m and 45cm long. His dad is going to extend it by 375cm. How long will it be then?

• •

4. To go camping for the weekend it costs £32.50p for the tent, £4 for the car, £4.50 for each adult and £2.60 for each child. How much will it cost 2 families with 2 parents and 2 children each to camp for the weekend if they go in their separate cars but share a tent?

• •

5. What three numbers could have a total of 572?

• •

6. The local bakery always donates its leftover bread to three local cafés. On Saturday they had 165 bread rolls, 149 chelsea buns and 70 french sticks left. If the bakery gives each café an equal total of items, how many did each café receive?

• •

7. What operation does the * represent?

25 * 15 = 375

How can you check?

• •

8. The pond in Trudy's garden holds 335 litres of water. In the summer 167 litres evaporate in the hot weather. How much water is left in the pond?

Reasoning about numbers or shapes

Whole class activity

How many squares can you see?

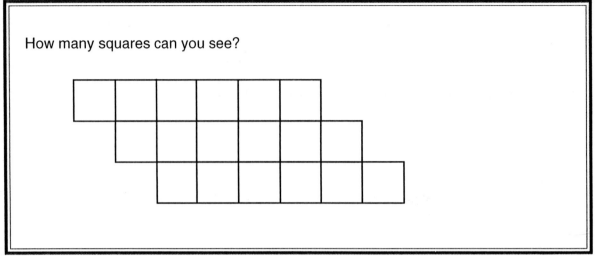

If the length of a cube is 5cm. How far will the perimeter around the net of the cube be?

Which letter of the alphabet has the most lines of symmetry?

1. Does the net of a triangular prism have more right angles than the net of a cuboid?

....

2. How many squares do you need to make a square twice this size?

....

3. What 2D shapes would you need to make a 3D model of a cube?.

....

4. Draw a shape with 2 lines of symmetry

....

5. 4 lines make a square. How many squares can you make with 8 lines?

....

6. Draw and name a 3D shape that has more than 4 edges.

....

7. What shape has more edges, a cube or a cylinder?

....

8. How many lines of symmetry are there in the letters of the word **WAIT**?

Lesson 1b

1. Does the net of a hexagonal prism have more right angles than the net of a pentagonal prism?

. . . .

2. How many squares do you need to make a square 4 times this size?

. . . .

3. What 2D shapes would you need to make a 3D model of a triangular prism?

. . . .

4. Draw a shape with 4 lines of symmetry.

. . . .

5. 4 lines make a square. How many squares can you make with 10 lines?

. . . .

6. Draw and name a 3D shape that has more than 6 edges.

. . . .

7. What shape has more edges, a pentagonal prism or a hexagonal prism?

. . . .

8. How many lines of symmetry are there in the letters of the word **WATCH**?

1. Does the net of a regular hexagonal prism have more right angles than the net of a regular heptagonal prism?

....

2. How many squares do you need to make a square 8 times this size?

....

3. What 2D shapes would you need to make a 3D model of a regular hexagonal prism?

....

4. Draw a shape with more than 4 lines of symmetry.

....

5. 4 lines make a square. How many squares can you make with 15 lines?

....

6. Draw and name a 3D shape that has more than 8 edges.

....

7. Which has more edges, a pentagonal prism or an octagonal prism?

....

8. How many lines of symmetry are there in the letters of the word **MOTHER**?

Write down an example that supports this statement.
'A multiple of 8 is both a multiple of 4 and a multiple of 2.'

Explain how you would do this calculation:

5008 – 4993

Explain how you would calculate the area of a rectangle.

Lesson
2a

1. Write down an example to match this statement.
'The size of any angle on a straight line will be 180° minus the other angle.'

• •

2. Find 2 consecutive numbers with a product of 110.

• •

3. Explain how you would do this calculation:

138 ÷ 23

• •

4. Write down an example that supports this statement.
'If you subtract a larger number from a smaller one you get a negative answer.'

• •

5. What 2-digit number, which is a multiple of 5, has a tens digit number and a units digit number, that when added together make 3?

• •

6. Explain how you would do this calculation:

13 x 12

• •

7. Write down an example that supports this statement.
'The area of a square is the length of a side multiplied by itself.'

• •

8. Find 3 ways to complete.

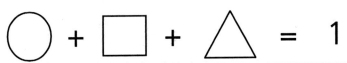

Reasoning about numbers or shapes

Lesson 2b

1. Write down an example to match this statement.
'The size of an angle inside a quadrilateral will be 360° minus the sum of the other 3 angles.'

· ·

2. Find 2 consecutive numbers with a product of 240.

· ·

3. Explain how you would do this calculation:

253 ÷ 23

· ·

4. Write down an example that supports this statement.
'If you subtract a larger number from a smaller one you get a negative answer.'

· ·

5. What 2-digit number which is a multiple of 8, has a tens digit number that is half the units digit number?

· ·

6. Explain how you would do this calculation:

18 x 12

· ·

7. Write down an example that supports this statement.
'The area of a rectangle is its length multiplied by its width.

· ·

8. Find 3 ways to complete.

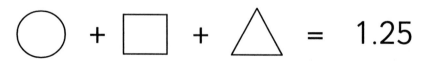

1. Write down an example to match this statement.
'The size of any angle in a pentagon will be 540° minus the sum of the other 4 angles.'

• •

2. Find 2 consecutive numbers with a product of 342.

• •

3. Explain how you would do this calculation:

$483 \div 23$

• •

4. Write down an example that supports this statement.
'If you subtract a larger number from a smaller one you get a negative answer.'

• •

5. What 2-digit number which is a multiple of 12, has a tens digit number that is 3 more than the units digit number?

• •

6. Explain how you would do this calculation:

28×12

• •

7. Write down an example that supports this statement: The area of a right angled triangle is half the length multiplied by the width.

• •

8. Find 3 ways to complete.

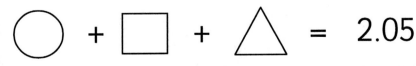

$$\bigcirc + \square + \triangle = 2.05$$

Reasoning about numbers or shapes

Whole class activity

If this shape is two thirds of the pattern what could the whole pattern look like?

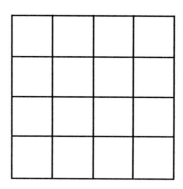

Split this shape into two equal areas. Can it be done in more than 5 ways?

If the perimeter of a pentagon is 50cm long, how long is each side?

1. How many more faces does a pentagonal prism have than a cuboid?

2. How many cubes would you need to make another cube which is 4 cubes wide?

3. What shapes would you need to make a 3D model of a triangular based pyramid?

4. You have a square piece of paper. What is the least number of cuts you have to make to turn the square into a pentagon?

5. Draw and name a 3D shape that has more than 4 faces.

6. Draw five squares in the shape of a cross. How many lines of symmetry are there? Draw them on.

7. What could the time be if the hands on an analogue clock show a right angle?

8. How many different shaped rectangles can you make with 12 square blocks?

Lesson 3b

1. How many more faces does a heptagonal prism have than a hexagonal prism?

· ·

2. How many cubes do you need to make another cube that is 8 cubes wide?

· ·

3. What shapes would you need to make a 3D model of a square based pyramid?

· ·

4. You have a square piece of paper. What is the least number of cuts you have to make to turn the square into a hexagon?

· ·

5. Draw and name a 3D shape that has more than 6 faces.

· ·

6. Draw nine squares in the shape of a cross. How many lines of symmetry are there? Draw them on.

· ·

7. What could the time be if the hands on an analogue clock show a 180º angle?

· ·

8. How many different shaped rectangles can you make with 18 square blocks?

1. How many more faces does an octagonal prism have than a hexagonal prism?

   •

2. How many cubes do you need to make another cube that is 10 cubes wide?

   •

3. Which shapes would you need to make a 3D model of a pentagonal based pyramid?

   •

4. You have a square piece of paper. What is the least number of cuts you have to make to turn the square into a octagon?

   •

5. Draw and name a 3D shape that has more than 7 faces.

   •

6. Draw 13 squares in the shape of a cross. How many lines of symmetry are there? Draw them on.

   •

7. What could the time be if the hands on an analogue clock show a 60º angle?

   •

8. How many different shaped rectangles can you make with 24 square blocks?

Reasoning about numbers or shapes

Whole class activity

Find 3 consecutive numbers with a total of 198.

Explain how you would do this calculation:

$\frac{1}{6}$ of 318

Write down an example that supports this statement.

'If you multiply any whole number by 100 you move it two digits to the right.'

---

www.brilliantpublications.co.uk
Maths Problem Solving – Year 5

1. What could the missing digits be?

   5** ÷ ** = 32

   ●

2. Write down an example that supports this statement.
   'The product of two numbers where one ends in 0 will result in an answer that ends in 0.'

   ●

3. What three numbers could you add to make 500?

   ●

4. Write down an example that supports this statement.
   'A regular shape with more than 4 sides has more than 4 lines of symmetry.'

   ●

5. Explain how you would do this calculation:

   200 x 50

   ●

6. The rule is 'Double the next consecutive number and add 3'. Start on 1. What will the next 6 numbers be?

   ●

7. Write down an example that supports this statement.
   'If you add 2 even and 2 odd numbers together you get an even answer.'

   ●

8. Explain how you would do this calculation:

   29 x 40

Reasoning about numbers or shapes

What could the missing digits be?

1.   8** ÷ ** = 32

....

2.   Write down an example that supports this statement.
'The product of three numbers where one ends in 0 will result in an answer that ends in 0.'

....

3.   What three numbers could you add to make 1000?

....

4.   Write down an example that supports this statement.
'A regular shape with more than 5 sides has more than 5 lines of symmetry.'

....

5.   Explain how you would do this calculation:

     600 x 50

....

6.   The rule is 'Double the next consecutive number and add 5'. Start on 1. What will the next 6 numbers be?

....

7.   Write down an example that supports this statement.
'If you add 3 even and 3 odd numbers together you get an odd answer.'

....

8.   Explain how you would do this calculation:

     59 x 40

1. What could the missing digits be?

   9** ÷ ** = 32

   •

2. Write down an example that supports this statement.
   'The product of four numbers where one ends in 0 will result in an answer that ends in 0.'

   •

3. What four numbers could you add to make 1000?

   •

4. Write down an example that supports this statement.
   'A regular shape with more than 6 sides has more than 6 lines of symmetry.'

   •

5. Explain how you would do this calculation:

   800 x 50

   •

6. The rule is 'Triple the next consecutive number and add 5'. Start on 1. What will the next 6 numbers be?

   •

7. Write down an example that supports this statement.
   'If you add 4 even and 4 odd numbers together you get an even answer.'

   •

8. Explain how you would do this calculation:

   159 x 40

This page may be photocopied by the purchasing institution only.

© Catherine Yemm

www.brilliantpublications.co.uk

Maths Problem Solving – Year 5   49

Reasoning about numbers or shapes

Whole class activity

What could the time be if the hands on an analogue clock show an angle of 270º?

Which has more edges, a square-based pyramid or a triangular prism?

If an umbrella has 8 spokes, what shape is it? How many lines of symmetry does it have?

1. Join together a square and a pentagon. How many sides does the new shape have?

....

2. Which shapes would you need to make a 3D model of a cylinder?

....

3. How many right angles does the net of a cuboid have?

....

4. Draw three different pentominoes.

....

5. Write your initials in capital letters. Draw on all the lines of symmetry in the letters.

....

6. Draw an irregular pentagon.

....

7. How many cubes would you need to make a cuboid that is 4 cubes wide, 2 cubes high and 8 cubes long?

....

8. Draw and name a 3D shape that has more than 6 right angles.

Lesson 5b

1. Join together a square and a hexagon. How many sides does the new shape have?

....

2. Which shapes would you need to make a 3D model of a hexagonal prism?

....

3. How many right angles does the net of a triangular prism have?

....

4. Draw four different pentominoes.

....

5. Write your first name in capital letters. Draw on all the lines of symmetry in the letters.

....

6. Draw an irregular hexagon.

....

7. How many cubes would you need to make a cuboid that is 8 cubes wide, 2 cubes high and 12 cubes long?

....

8. Draw and name a 3D shape that has more than 8 right angles.

Lesson
5C

1. Join together a heptagon and a hexagon. How many sides does the new shape have?

• •

2. Which shapes would you need to make a 3D model of an octagonal prism?

• •

3. How many right angles does the net of an pentagonal prism have?

• •

4. Draw five different pentominoes.

• •

5. Write your first name and surname in capital letters. Draw on all the lines of symmetry in the letters.

• •

6. Draw an irregular octagon.

• •

7. How many cubes would you need to make a cuboid that is 12 cubes wide, 4 cubes high and 16 cubes long?

• •

8. Draw and name a 3D shape that has more than 10 right angles.

This page may be photocopied by the purchasing institution only.

© Catherine Yemm

www.brilliantpublications.co.uk

Maths Problem Solving – Year 5   53

Reasoning about numbers or shapes

Whole class activity

Explain how you would do this calculation:

7% of 250

Write down an example that supports this statement.
'If you add two consecutive numbers the answer will be double the first number plus 1.'

Write the next two numbers in the sequence.

8   15   29   57   113   _____   _____

---

1. Find two consecutive numbers that add up to 175.

· ·

2. Explain how you would do this calculation:

   19 x 40

· ·

3. Explain how to find the number of seconds in *n* number of minutes.

· ·

4. The angles inside a triangle add up to 360º.

   True

   False

· ·

5. What three-digit number is a multiple of 11 where the three digits add up to 11?

· ·

6. Write down an example that supports this statement.
   'Multiples of 7 can be odd or even.'

· ·

7. Explain how you would do this calculation:

   156 + 163

· ·

8. Write down an example that supports this statement.
   'If you double an even number 3 times, the answer will be even.'

Lesson 6b

1. Find two consecutive numbers that add up to 275.

• •

2. Explain how you would do this calculation:

39 x 40

• •

3. Explain how to find the number of seconds in *n* number of hours.

• •

4. The angles inside a pentagon add up to 360°.

True

False

• •

5. What three-digit number is a multiple of 12 where the three digits add up to 12?

• •

6. Write down an example that supports this statement.
'Multiples of 9 can be odd or even.'

• •

7. Explain how you would do this calculation:

456 + 163

• •

8. Write down an example that supports this statement.
'If you double an even number 5 times, the answer will be even.'

1. Find 2 consecutive numbers that add up to 475.

● ●

2. Explain how you would do this calculation:

69 x 40

● ●

3. Explain how to find the number of seconds in *n* number of days.

● ●

4. The angles inside an octagon add up to 360°.

True

False

● ●

5. What three-digit number is a multiple of 15 where the three digits add up to 15?

● ●

6. Write down an example that supports this statement.
'Multiples of 13 can be either odd or even.'

● ●

7. Explain how you would do this calculation:

456 + 365

● ●

8. Write down an example that supports this statement.
'If you double an even number 7 times, the answer will be even.'

Problems involving 'real life', money or measures

I think of a number, then divide it by 20. My answer is 25. What was my number?

Paul bought some items costing £3.27, £14.92, 69p, £2.93 and 58p. How much was the total?

Mrs Hardy fills her car up on Monday morning with 60 litres of petrol. Her car uses up 7 litres of petrol every day taking her children to school and she uses 9 litres of petrol on Saturday and 8 litres on Sunday. How much petrol does she have left on Sunday evening?

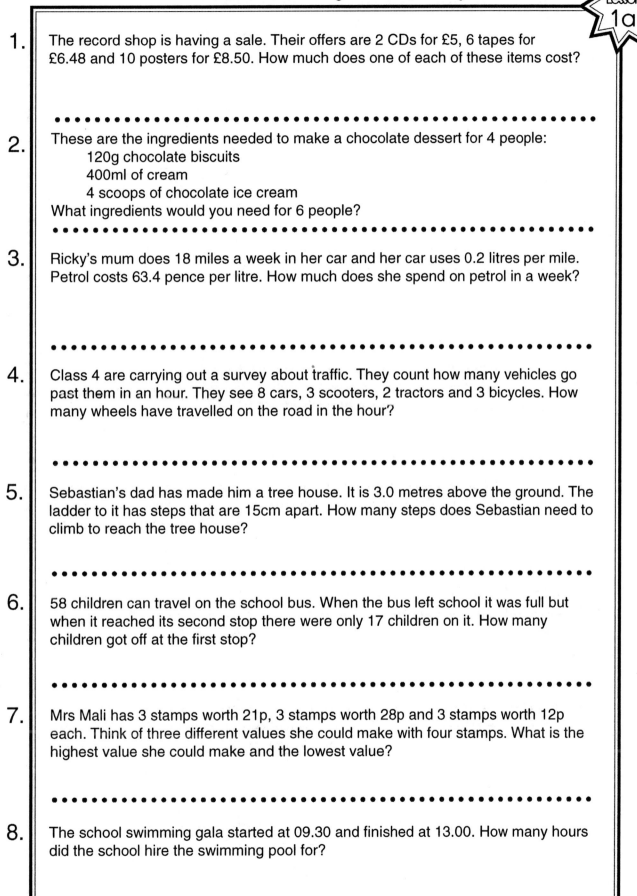

1. The record shop is having a sale. Their offers are 2 CDs for £5, 6 tapes for £6.48 and 10 posters for £8.50. How much does one of each of these items cost?

2. These are the ingredients needed to make a chocolate dessert for 4 people:
   120g chocolate biscuits
   400ml of cream
   4 scoops of chocolate ice cream
   What ingredients would you need for 6 people?

3. Ricky's mum does 18 miles a week in her car and her car uses 0.2 litres per mile. Petrol costs 63.4 pence per litre. How much does she spend on petrol in a week?

4. Class 4 are carrying out a survey about traffic. They count how many vehicles go past them in an hour. They see 8 cars, 3 scooters, 2 tractors and 3 bicycles. How many wheels have travelled on the road in the hour?

5. Sebastian's dad has made him a tree house. It is 3.0 metres above the ground. The ladder to it has steps that are 15cm apart. How many steps does Sebastian need to climb to reach the tree house?

6. 58 children can travel on the school bus. When the bus left school it was full but when it reached its second stop there were only 17 children on it. How many children got off at the first stop?

7. Mrs Mali has 3 stamps worth 21p, 3 stamps worth 28p and 3 stamps worth 12p each. Think of three different values she could make with four stamps. What is the highest value she could make and the lowest value?

8. The school swimming gala started at 09.30 and finished at 13.00. How many hours did the school hire the swimming pool for?

This page may be photocopied by the purchasing institution only.

© Catherine Yemm

www.brilliantpublications.co.uk

Maths Problem Solving – Year 5   59

Lesson 1b

1. The record shop is having a sale. Their offers are 4 CDs for £5, 6 tapes for £12.48 and 10 posters for £17.50. How much does one of each of these items cost?

2. These are the ingredients needed to make a chocolate dessert for 4 people:
   150g chocolate biscuits
   500ml of cream
   6 scoops of chocolate ice cream
   What ingredients would you need for 6 people?

3. Ricky's mum does 32 miles a week in her car and her car uses 0.2 litres per mile. Petrol costs 63.4 pence per litre. How much does she spend on petrol in a week?

4. Class 4 are carrying out a survey about traffic. They count how many vehicles go past them in an hour. They see 14 cars, 7 scooters, 2 tractors and 6 bicycles. How many wheels have travelled on the road in the hour?

5. Sebastian's dad has made him a tree house. It is 4.35 metres above the ground. The ladder to it has steps that are 15cm apart. How many steps does Sebastian need to climb to reach the tree house?

6. 62 children can travel on the school bus. When the bus left school it was full but when it reached its second stop there were only 17 children on it. How many children got off at the first stop?

7. Mrs Mali has 3 stamps worth 27p, 3 stamps worth 31p and 3 stamps worth 16p each? Think of three different values she could make with four stamps. What is the highest value she could make and the lowest value?

8. The school swimming gala started at 09.30 and finished at 15.45. How many hours did the school hire the swimming pool for?

1. The record shop is having a sale. Their offers are 2 CDs for £15, 6 tapes for £26.46 and 10 posters for £18.50. How much does one of each of these items cost?

• •

2. These are the ingredients needed to make a chocolate dessert for 4 people.
   280g chocolate biscuits
   700ml of cream
   8 scoops of chocolate ice cream
What ingredients would you need for 6 people?

• •

3. Ricky's mum does 62 miles a week in her car and her car uses 0.2 litres per mile. Petrol costs 63.4 pence per litre. How much does she spend on petrol in a week?

• •

4. Class 4 are carrying out a survey about traffic. They count how many vehicles go past them in an hour. They see 24 cars, 17 scooters, 2 tractors and 16 bicycles. How many wheels have travelled on the road in the hour?

• •

5. Sebastian's dad has made him a tree house. It is 6.45 metres above the ground. The ladder to it has steps that are 15cm apart. How many steps does Sebastian need to climb to reach the tree house?

• •

6. 81 children can travel on the school bus. When the bus left school it was full but when it reached its second stop there were only 24 children on it. How many children got off at the first stop?

• •

7. Mrs Mali has 3 stamps worth 47p, 3 stamps worth 38p and 3 stamps worth 26p each. Think of three different values she could make with four stamps. What is the highest value she could make and the lowest value?

• •

8. The school swimming gala started at 09.30 and finished at 17.25. How many hours did the school hire the swimming pool for?

Problems involving 'real life', money or measures

Mr Walker bought 5kg of coconuts for the coconut shy at the school fête. Each coconut weighed 250g. 12 children won a coconut at the fête. How heavy was the bag of coconuts Mr Walker carried home?

Jessica's library book is 324 pages long. She is 22 pages away from reaching half way. How many pages does she have left to read?

The blue relay team on sports day ran the following times; child 1 – 45.2 seconds, child 2– 68.5 seconds, child 3 – 84.3 seconds, child 4 – 71.3 seconds. How long did it take the team to complete their race, in minutes and seconds?

1. Amy has decided to give 10p a week to charity. After 2 years, how much will she have donated?

. . . .

2. Jay started making his Halloween costume on September 20th and finished it the day before Halloween. How many days did it take him to make it?

. . . .

3. To make 1 litre of potato soup Brian uses 8 potatoes. How many potatoes will he need if he wants to make 2500 millilitres of soup?

. . . .

4. When Elizabeth was 4 years old she was 100cm tall. A year later she was 102cm tall. If she keeps on growing at the same rate how tall will she be in metres when she is 11 years old?

. . . .

5. I think of a number and subtract 18 and divide by 4. The answer is 25. What was my number?

. . . .

6. Liana goes to the shop and spends £15.70 on fruit and vegetables. How much change will she have from £20? What coins and notes could she be given?

. . . .

7. For her birthday Polly's dad paid for her and 3 friends to go bowling. He paid for the children to have two games. The total bill was £18.40. How much did it cost for one person to have one game?

. . . .

8. When shopping Rachel noticed some special offers. The 1 litre bottle of lemonade she usually bought had 25% extra in it for free. How much lemonade was in the bottle?

www.brilliantpublications.co.uk

Lesson 2b

1. Amy has decided to give 18p a week to charity. After 2 years, how much will she have donated?

• •

2. Jay started making his Halloween costume on August 24th and finished it the day before Halloween. How many days did it take him to make it?

• •

3. To make 1 litre of potato soup Brian uses 8 potatoes. How many potatoes will he need if he wants to make 7250 millilitres of soup?

• •

4. When Elizabeth was 4 years old she was 100cm tall. A year later she was 104cm tall. If she keeps on growing at the same rate how tall will she be in metres when she is 11 years old?

• •

5. I think of a number and subtract 18 and divide by 8. The answer is 25. What was my number?

• •

6. Liana goes to the shop and spends £12.73 on fruit and vegetables. How much change will she have from £20? What coins and notes could she be given?

• •

7. For her birthday Polly's dad paid for her and 3 friends to go bowling. He paid for the children to have two games. The total bill was £18.80. How much did it cost for one person to have one game?

• •

8. When shopping Rachel noticed some special offers. The 2 litre bottle of lemonade she usually bought had 25% extra in it for free. How much lemonade was in the bottle?

www.brilliantpublications.co.uk

Lesson
2c

1. Amy has decided to give 28p a week to charity. After 2 years how much would she have donated?

....

2. Jay started making his Halloween costume on July 7th and finished it the day before Halloween. How many days did it take him to make it?

....

3. To make 1 litre of potato soup Brian uses 8 potatoes. How many potatoes will he need if he wants to make 9750 millilitres of soup?

....

4. When Elizabeth was 4 years old she was 100cm tall. A year later she was 105cm tall. If she keeps on growing at the same rate how tall will she be in metres when she is 18 years old?

....

5. I think of a number and subtract 38 and divide by 12. The answer is 25. What was my number?

....

6. Liana goes to the shop and spends £31.73 on fruit and vegetables. How much change will she have from £50? What coins and notes could she be given?

....

7. For her birthday Polly's dad paid for her and 3 friends to go bowling. He paid for the children to have two games. The total bill was £19.76. How much did it cost for one person to have one game?

....

8. When shopping Rachel noticed some special offers. The 3.5 litre bottle of lemonade she usually bought had 25% extra in it for free. How much lemonade was in the bottle?

Problems involving 'real life', money or measures

Whole class activity

The perimeter of the fence around the school playing field measures 210 metres. If the shape of the field is a rectangle and the widths are 28 metres each. How long is the field?

James is having a party for his birthday. His mum has made 72 sandwiches, 45 sausage rolls and 36 fairy cakes. After the party, a fifth of the sausage rolls are left, a quarter of the sandwiches are left and one third of the cakes are left. How much of each type of food was eaten?

If the exchange rate is 1.4 euros or 1.7 US dollars for £1, how many pounds does Alex have altogether if he has 27.2 US dollars and 53.2 euros?

Lesson
3a

1. It is 5 miles from Sheila's house to school and 8 miles from school to where her mum works. How many miles in a week does Sheila's mum travel if she drops her daughter off at school each morning on her way to work and then picks her up after school to take her home?

· ·

2. I think of a number, I double it then add 25. The answer is 81. What was my number?

· ·

3. Each video in the library costs £1.05 to hire. If Jerry hires a video every day for a fortnight, how much money will he spend?

· ·

4. Half of the flowers in Sheba's garden are yellow and half are white. If 47 are yellow, how many flowers does she have in her garden?

· ·

5. Jill is buying some sugar for the school canteen, but she cannot put more than 10kg in her trolley. A bag of sugar weighs 2.2kg. How many bags of sugar could Jill fit in her trolley?

· ·

6. Sasha went shopping in town and queued from 10.30 to 10.42 in the first shop, from 12.45 to 13.00 in the second shop and from 15.10 to 16.35 at the bus stop. How long altogether did she spend queuing?

· ·

7. The school secretary works from 9 o'clock until 3 o'clock. In this time the phone rings 70 times. Two quarters of the calls are from parents. How many calls are not from parents?

· ·

8. Tano has six shelves of books in his bedroom. Four of the shelves hold 12 books each and two of the shelves hold 27 books each. How many books does Tano have in his bedroom?

© Catherine Yemm

www.brilliantpublications.co.uk

Lesson 3b

1. It is 7 miles from Sheila's house to school and 12 miles from school to where her mum works. How many miles in a week does Sheila's mum travel if she drops her daughter off at school each morning on her way to work and then picks her up after school to take her home?

• •

2. I think of a number, I double it then add 45. The answer is 81. What was my number?

• •

3. Each video in the library costs £1.25 to hire. If Jerry hires a video every day for a fortnight, how much money will he spend?

• •

4. Half of the flowers in Sheba's garden are yellow and half are white. If 87 are yellow, how many flowers does she have in her garden?

• •

5. Jill is buying some sugar for the school canteen, but she cannot put more than 25kg in her trolley. A bag of sugar weighs 2.2kg. How many bags of sugar could Jill fit in her trolley?

• •

6. Sasha went shopping in town and queued from 10.30 to 10.47 in the first shop, from 12.45 to 13.07 in the second shop and from 15.11 to 16.36 at the bus stop. How long altogether did she spend queuing?

• •

7. The school secretary works from 9 o'clock until 3 o'clock. In this time the phone rings 95 times. Two fifths of the calls are from parents. How many calls are not from parents?

• •

8. Tano has six shelves of books in his bedroom. Four of the shelves hold 22 books each and two of the shelves hold 37 books each. How many books does Tano have in his bedroom?

1. It is 9 miles from Sheila's house to school and 15 miles from school to where her mum works. How many miles in a week does Sheila's mum travel if she drops her daughter off at school each morning on her way to work and then picks her up after school to take her home?

● ●

2. I think of a number, I double it then add 75. The answer is 167. What was my number?

● ●

3. Each video in the library costs £2.15 to hire. If Jerry hires a video every day for a fortnight, how much money will he spend?

● ●

4. Half of the flowers in Sheba's garden are yellow and half are white. If 127 are yellow, how many flowers does she have in her garden?

● ●

5. Jill is buying some sugar for the school canteen, but she cannot put more than 45kg in her trolley. A bag of sugar weighs 2.2kg. How many bags of sugar could Jill fit in her trolley?

● ●

6. Sasha went shopping in town and queued from 10.30 to 10.51 in the first shop, from 12.45 to 13.17 in the second shop and from 15.11 to 16.56 at the bus stop. How long altogether did she spend queuing?

● ●

7. The school secretary works from 9 o'clock until 3 o'clock. In this time the phone rings 105 times. Four fifths of the calls are from parents. How many calls are not from parents?

● ●

8. Tano has six shelves of books in his bedroom. Four of the shelves hold 32 books each and two of the shelves hold 57 books each. How many books does Tano have in his bedroom?

Problems involving 'real life', money or measures

The Fletcher family sat down at 14:15 to watch the home video of their holiday. They watched 55 minutes of it, then they had a 10 minute tea break, then they watched the next 1 hour and 37 minutes. What time did the video end?

At the beginning of Year 3 the class teacher weighed everybody. Tamsin weighed 32kg and Felix weighed 45kg. At the end of Year 4 they were weighed together and they weighed 90kg. If Felix put on 5kg more weight than Tamsin, how heavy were Tamsin and Felix at the end of Year 4?

Ribbon is on offer at the shop: if you buy 5 metres you get 25% extra free. If Lucy's mum buys 5 metres of red ribbon and 8 metres of yellow ribbon, how much free ribbon will she get altogether?

1. Freddie's parents are thinking of going on holiday to Egypt, Hong Kong or Jamaica. He has £10 spending money and the exchange rates are £1 = 10.4 Egyptian pounds, £1 = 98.6 Jamaican dollars, £1 = 13.3 Hong Kong dollars. How much spending money will he have at each destination?

2. At the school bonfire night display some of the parents are selling hot chocolate. Each mug holds 220ml of hot chocolate. How many litres of hot chocolate will the parents have to make if they have sold 20 tickets.

3. Bessie went to bed at 5.45 pm on Tuesday evening and slept right through until 7.50 am on Wednesday morning. How much sleep did she get?

4. Mrs Miller has bought some lollies and sweets to put in goodie bags to sell at the school fête. She has bought 64 lollies and 56 sweets. If she puts 4 in each bag, how many bags will she make?

5. Bali rides his bike at 15km an hour from his house to the village post office. It takes him 15 minutes to get there. How far will he have cycled if he goes to the post office and back?

6. At the school fête it costs £1.10 to buy a strip of raffle tickets and 55p to have a go on the coconut shy. Leigh has £4.50 to spend. What is the largest quantity of raffle tickets he could buy? What is the most number of goes he could have on the coconut shy?

7. All of the classes in the school are having a sponsored spelling test to raise money. Classes 1 and 2 have 10 words to spell, classes 3 and 4 have 15 words to spell, classes 5 and 6 have 20 words to spell. If there are 26 children in each class, how many words will they write down between them?

8. Paula's dad has put down a deposit on a new tent for her at the camping shop. The tent costs £195 and he has put down a 10% deposit. How much has he already paid? How much does he still have to pay before he can take the tent home?

© Catherine Yemm

Lesson 4b

1. Freddie's parents are thinking of going on holiday to Egypt, Hong Kong or Jamaica. He has £20 spending money and the exchange rates are £1 = 10.4 Egyptian pounds, £1 = 98.6 Jamaican dollars, £1 = 13.3 Hong Kong dollars. How much spending money will he have at each destination?

• •

2. At the school bonfire night display some of the parents are selling hot chocolate. Each mug holds 220ml of hot chocolate. How many litres of hot chocolate will the parents have to make if they have sold 50 tickets?

• •

3. Bessie went to bed at 5.45 pm on Tuesday evening and slept right through until 7.20 am on Wednesday morning. How much sleep did she get?

• •

4. Mrs Miller has bought some lollies and sweets to put in goodie bags to sell at the school fête. She has bought 124 lollies and 72 sweets. If she puts 4 in each bag, how many bags will she make?

• •

5. Bali rides his bike at 32km an hour from his house to the village post office. It takes him 15 minutes to get there. How far will he have cycled if he goes to the post office and back?

• •

6. At the school fête it costs £1.55 to buy a strip of raffle tickets and 75p to have a go on the coconut shy. Leigh has £6.50 to spend. What is the largest quantity of raffle tickets he could buy? What is the most number of goes he could have on the coconut shy?

• •

7. All of the classes in the school are having a sponsored spelling test to raise money. Classes 1 and 2 have 10 words to spell, classes 3 and 4 have 25 words to spell, classes 5 and 6 have 50 words to spell. If there are 26 children in each class, how many words will they write down between them?

• •

8. Paula's dad has put down a deposit on a new tent for her at the camping shop. The tent costs £195 and he has put down a 20% deposit. How much has he already paid? How much does he still have to pay before he can take the tent home?

1. Freddie's parents are thinking of going on holiday to Egypt, Hong Kong or Jamaica. He has £35 spending money and the exchange rates are £1 to 10.4 Egyptian pounds, £1 to 98.6 Jamaican dollars, £1 to 13.3 Hong Kong dollars. How much spending money will he have at each destination?

2. At the school bonfire night display some of the parents are selling hot chocolate. Each mug holds 220ml of hot chocolate. How many litres of hot chocolate will the parents have to make if they have sold 75 tickets?

3. Bessie went to bed at 5.42 pm on Tuesday evening and slept right through until 7.37 am on Wednesday morning. How much sleep did she get?

4. Mrs Miller has bought some lollies and sweets to put in goodie bags to sell at the school fête. She has bought 248 lollies and 144 sweets. If she puts 4 in each bag, how many bags will she make?

5. Bali rides his bike at 45km an hour from his house to the village post office. It takes him 15 minutes to get there. How far will he have cycled if he goes to the post office and back?

6. At the school fête it costs £1.55 to buy a strip of raffle tickets and 75p to have a go on the coconut shy. Leigh has £12.50 to spend. What is the largest quantity of raffle tickets he could buy? What is the most number of goes he could have on the coconut shy?

7. All of the classes in the school are having a sponsored spelling test to raise some money classes 1 and 2 have 15 words to spell, classes 3 and 4 have 30 words to spell, classes 5 and 6 have 60 words to spell. If there are 26 children in each class, how many words will they write down between them?

8. Paula's dad has put down a deposit on a new tent for her at the camping shop. The tent costs £195 and he has put down a 30% deposit. How much has he already paid? How much does he still have to pay before he can take the tent home?

Problems involving 'real life', money or measures

Whole class activity

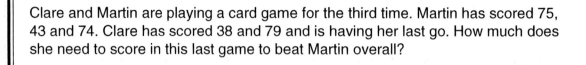

Clare and Martin are playing a card game for the third time. Martin has scored 75, 43 and 74. Clare has scored 38 and 79 and is having her last go. How much does she need to score in this last game to beat Martin overall?

Everyone in Mrs White's street pays 17p per litre for their water. Mrs White's bill last month was £8.50. The water company have just put the cost of water up to 19p per litre. If she uses the same amount of water per month, what will her bill be this time?

Jack is buying some books in the sale. The Science books have 25% off and the Sport books have 30% off. He chooses a book about swimming that was £12 and a book about dissolving that was £22. How much does he need to pay the shopkeeper after the discounts have been taken off?

Lesson
5a

1. Keira uses 20 centimetres of wool to knit a tenth of a woollen scarf. How many metres of wool would she need to make 3 scarves?

....

2. Tickets to the cinema cost £2.50 for an adult and £1.65 for a child. How much would it cost Mr and Mrs Weaver and their triplets to go?

....

3. The school's Christmas fête is having a 'guess the weight of the cake' competition. Judith guessed 8.2kg but it only weighed a quarter of that. How many kilograms did the cake weigh?

....

4. Jamie's mum put his trousers in the washing machine and they shrank. They used to measure 110cm in length but they have lost 10% of their length. How long are they now?

....

5. School starts at 08.50. Stuart arrived at school at 08.10, Toby and Janet arrived 10 minutes later. Carol arrived 15 minutes after that and Simon arrived 6 minutes later. What time did Simon arrive? Was anyone late for school?

....

6. Dylan has been saving his pocket money for the last year to buy a new bike. For the first 3 months of the year he saved £1.50 per month, for the next 4 months he saved £2.75 per month and for the last 5 months he saved £2.05 per month. How much has he saved in the year?

....

7. Leanne has a job as a paper girl. On Sunday morning she delivers 16 papers and number 54 is the last delivery in her street. At what house number did she start, if everyone before had been given a paper?

....

8. The school cook made 100 hot cross buns for lunch. Year 1 children ate one tenth of them, Year 2 ate half of them and Year 3 ate two tenths of them. How many were left when Year 4 arrived for their lunch?

Lesson 5b

1. Keira uses 40 centimetres of wool to knit a tenth of a woollen scarf. How many metres of wool would she need to make 5 scarves?

· ·

2. Tickets to the cinema cost £3.75 for an adult and £2.65 for a child. How much would it cost Mr and Mrs Weaver and their triplets to go?

· ·

3. The school's Christmas fête is having a 'guess the weight of the cake' competition. Judith guessed 12.8kg but it only weighed a quarter of that. How many kilograms did the cake weigh?

· ·

4. Jamie's mum put his trousers in the washing machine and they shrank. They used to measure 110cm in length but they have lost 20% of their length. How long are they now?

· ·

5. School starts at 08.50. Stuart arrived at school at 08.17, Toby and Janet arrived 9 minutes later. Carol arrived 15 minutes after that and Simon arrived 6 minutes later. What time did Simon arrive? Was anyone late for school?

· ·

6. Dylan has been saving his pocket money for the last year to buy a new bike. For the first 3 months of the year he saved £3.50 per month, for the next 4 months he saved £4.75 per month and for the last 5 months he saved £5.05 per month. How much has he saved in the year?

· ·

7. Leanne has a job as a paper girl. On Sunday morning she delivers 26 papers and number 74 is the last delivery in her street. At what house number did she start, if everyone before had been given a paper?

· ·

8. The school cook made 150 hot cross buns for lunch. Year 1 children ate one tenth of them, Year 2 ate half of them and Year 3 ate two tenths of them. How many were left when Year 4 arrived for their lunch?

1. Keira uses 35 centimetres of wool to knit a tenth of a woollen scarf. How many metres of wool would she need to make 7 scarves?

• •

2. Tickets to the cinema cost £4.75 for an adult and £3.65 for a child. How much would it cost Mr and Mrs Weaver and their triplets to go?

• •

3. The school's Christmas fête is having a 'guess the weight of the cake' competition. Judith guessed 24.8kg but it only weighed a quarter of that. How many kilograms did the cake weigh?

• •

4. Jamie's mum put his trousers in the washing machine and they shrank. They used to measure 110cm in length but they have lost 35% of their length. How long are they now?

• •

5. School starts at 08.50. Stuart arrived at school at 08.07, Toby and Janet arrived 19 minutes later. Carol arrived 15 minutes after that and Simon arrived 6 minutes later. What time did Simon arrive? Was anyone late for school?

• •

6. Dylan has been saving his pocket money for the last year to buy a new bike. For the first 3 months of the year he saved £4.65 per month, for the next 4 months he saved £5.75 per month and for the last 5 months he saved £6.15 per month. How much has he saved in the year?

• •

7. Leanne has a job as a paper girl. On Sunday morning she delivers 48 papers and house number 94 is the last delivery in her street. At what house number did she start, if everyone before had been given a paper?

• •

8. The school cook made 280 hot cross buns for lunch. Year 1 children ate one tenth of them, Year 2 ate half of them and Year 3 ate two tenths of them. How many were left when Year 4 arrived for their lunch?

Problems involving 'real life', money or measures

Whole class activity

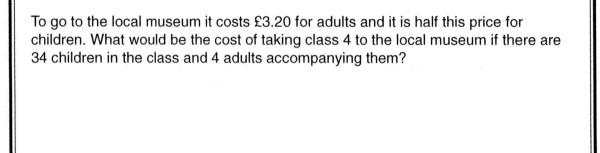

To go to the local museum it costs £3.20 for adults and it is half this price for children. What would be the cost of taking class 4 to the local museum if there are 34 children in the class and 4 adults accompanying them?

A marathon is 26 miles long. Peter enters the marathon and runs one tenth of the way, then walks for a quarter of the way, then runs again for the rest of the way. How far did he run and how far did he walk altogether?

I think of a number. I multiply it by 15 then subtract 25. My answer is 425. What was my number?

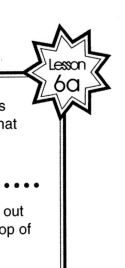

Lesson
6a

1. Mr Johns has an allotment that is 20 metres long by 30 metres wide. He wants to split it up into 4 smaller segments to grow 4 different types of vegetable. What area would each segment have?

2. Kieran enjoys painting and has decided to sell some of his pictures. He works out how much it cost him to buy the materials he needed and then adds 50% on top of that. If it cost him to £9.60 to make a picture, how much will he sell it for?

3. The school courtyard has a paved area of 32 square metres. The head teacher wants to lay some grass in one area of it. If a quarter of the courtyard becomes covered in grass, how much of the courtyard will still be paved? What length and width could the grassed area be?

4. It is a Saturday and Georgina's grandma is coming to stay in exactly 6 weeks time. How many week days will she have to wait until she sees her?

5. Veda was having a bath when the plug got stuck in the plug hole. She had to bail the water out with a 2.5 litre jug. If the bath had 30 litres of water in it, how many jugs of water did she have to bail out to empty the bath?

6. There are 12 chocolates in a box. The shopkeeper sells 13 boxes. How many chocolates has he sold?

7. The cost of a litre of orange juice is 37p. Joyce needs to buy 30 litres for her daughter's party. How much will she spend on orange juice? How much change will she have left from £20?

8. Danielle, Ella and Harriet collect stickers. Danielle has 132, Ella has 109 and Harriet has 143. How many do they have between them? How many more would they need to have 600 between them?

Lesson 6b

1. Mr Johns has an allotment that is 40 metres long by 60 metres wide. He wants to split it up into 4 smaller segments to grow 4 different types of vegetable. What area would each small field have?

2. Kieran enjoys painting and has decided to sell some of his pictures. He works out how much it cost him to buy the materials he needed and then adds 50% on top of that. If it cost him to £12.74 to make a picture, how much will he sell it for?

3. The school courtyard has a paved area of 56 square metres. The head teacher wants to lay some grass in one area of it. If a quarter of the courtyard becomes covered in grass, how much of the courtyard will still be paved? What length and width could the grassed area be?

4. It is a Saturday and Georgina's grandma is coming to stay in exactly 9 weeks time. How many week days will she have to wait until she sees her.

5. Veda was having a bath when the plug got stuck in the plug hole. She had to bail the water out with a 2.5 litre jug. If the bath had 70 litres of water in it, how many jugs of water did she have to bail out to empty the bath?

6. There are 15 chocolates in a box. The shopkeeper sells 13 boxes. How many chocolates has he sold?

7. The cost of a litre of orange juice is 57p. Joyce needs to buy 45 litres for her daughter's party. How much will she spend on orange juice? How much change will she have left from £30?

8. Danielle, Ella and Harriet collect stickers. Danielle has 232, Ella has 209 and Harriet has 243. How many do they have between them? How many more would they need to have 1000 between them?

1. Mr Johns has an allotment that is 52 metres long by 64 metres wide. He wants to split it up into 4 smaller segments to grow 4 different types of vegetable. What area would each small segment have?

2. Kieran enjoys painting and has decided to sell some of his pictures. He works out how much it cost him to buy the materials he needed and then adds 50% on top of that. If it cost him to £27.60 to make a picture, how much will he sell it for?

3. The school courtyard has a paved area of 86 square metres. The head teacher wants to lay some grass in one area of it. If a quarter of the courtyard becomes covered in grass, how much of the courtyard will still be paved? What length and width could the grassed area be?

4. It is a Saturday and Georgina's grandma is coming to stay in exactly 13 weeks time. How many week days will she have to wait until she sees her.

5. Veda was having a bath when the plug got stuck in the plug hole. She had to bail the water out with a 2.5 litre jug. If the bath had 135 litres of water in it, how many jugs of water did she have to bail out to empty the bath?

6. There are 25 chocolates in a box. The shopkeeper sells 13 boxes. How many chocolates has he sold?

7. The cost of a litre of orange juice is 87p. Joyce needs to buy 45 litres for her daughter's party. How much will she spend on orange juice? How much change will she have left from £50?

8. Danielle, Ella and Harriet collect stickers. Danielle has 432, Ella has 309 and Harriet has 543. How many do they have between them? How many more would they need to have 1500 between them?

Answer sheet
Making decisions

Lesson 1 (page 10)

A: story; B: minus; C: 22, 4 left over

Lessons 1a–1c (pages 11–13)

Q	1a	1b	1c
1	134	234	534
2	3 litres	4 litres	9 litres
3	41	71	21
4	1 hr, 20 mins.	3 hrs, 50 mins.	5 hrs, 46 mins.
5	Natalie 229 Ryan 240	Natalie 379 Ryan 390	Natalie 579 Ryan 690
6	62.25	101.25	148.5
7	12	16	14
8	16	66	216

Lesson 2 (page 14)

A: £150; B: 20:05; C: 153 litres

Lessons 2a–2c (pages 15–17)

Q	2a	2b	2c
1	132	222	258
2	207	238	268
3	84	132	228
4	story	story	story
5	32kg	64kg	112kg
6	70cm	125cm	225cm
7	multiply	multiply	multiply
8	£12	£31	£52

Lesson 3 (page 18)

A: 40; B: 159; C: the bag by £1.20

Lessons 3a–3c (pages 19–21)

Q	3a	3b	3c
1	6.40 pm	7.55 pm	8.05 pm
2	divide	divide	divide
3	6m^2	12m^2	20m^2
4	52	96	164
5	45.5kg	86.5kg	146.5kg
6	£42	£62	£162
7	story	story	story
8	180ml	320ml	540ml

Lesson 4 (page 22)

A: £15.00; B: 259; C: 25kg

Lessons 4a–4c (pages 23–25)

Q	4a	4b	4c
1	26 days	57 days	148 days
2	25	45	90
3	add	add	add
4	£3.10	£6.30	£7.95
5	statement	statement	statement
6	14 tsp	25 tsp	37.4 tsp
7	251	411	711
8	43 days	85 days	185 days

Lesson 5 (page26)

A: 19; B: 12 hours, 27 minutes;

C: 7th March, 30th October

Lessons 5a–5c (pages 27–29)

Q	5a	5b	5c
1	38	46	76
2	58	118	318
3	11.2kg	10.4kg	14.8kg
4	minus	minus	minus
5	−4.6°	−4.8°	−10.8°
6	324cm	432cm	648cm
7	statement	statement	statement
8	£2.85	£3.75	£11.10

Lesson 6 (page 30)
A: 2 minutes, 19 seconds; B: £29.55;
C: 1068

Lessons 6a–6c (pages 31–33)

Q	6a	6b	6c
1	£7.80	£19.80	£34.20
2	statement	statement	statement
3	353cm	553cm	720cm
4	£39.80	£67.80	£68.90
5	any three numbers with a total of		
	172	372	572
6	62	92	128
7	multiply	multiply	multiply
8	68 litres	136 litres	168 litres

Reasoning about numbers and shape

Lesson 1 (page 34)
A: 26; B: 70cm; C: O

Lessons 1a–1c (pages 35–37)

Q	1a	1b	1c
1	no	yes	no
2	16	64	256
3	6 squares	2 triangles	2 hexagons
		3 rectangles	6 rectangles
4	N/A	N/A	N/A
5	14	26	64
6	eg cube prism	eg sqare-based pyramid	eg cube, triang. prism
7	a cube	hex. prism	oct. prism
8	5	6	6+

Lesson 2 (page 38)
A: 2 x 4 = 8; B: minus calculation;
C: multiply the length by the width

Lessons 2a–2c (pages 39–41)

Q	2a	2b	2c
1	N/A	N/A	N/A
2	10; 11	15; 16	18; 19
3	child to show workings out		
4	Various answers		
5	30	24 or 48	96
6	child to show workings out		
7	children to show examples		
8	Various answers		

Lesson 3 (page 42)
A: any pattern using that shape; B: yes;
C: 10cm

Lessons 3a–3c (pages 43–45)

Q	3a	3b	3c
1	1	1	2
2	64	512	1000
3	4 triangles	1 square 4 triangles	1 pentagon 5 triangles
4	1	2	4
5	cuboid etc	pentagonal prism	hexagonal prism
6	4	4	4
7	3 or 9 o'clock	6 o'clock	2 or 10 o'clock
8	3	3	4

Lesson 4 (page 46)
A: 65, 66, 67; B: division – class discussion; C: no, statement to be discussed.

Lessons 4a–4c (pages 47–49)

Q	4a	4b	4c
1	512÷16=32 544÷17=32 576÷18=32	800÷25=32 832÷26=32 864÷27=32 896÷28=32	928÷29=32 960÷30=32 992÷31=32
2	20x4=80etc	20x3x11=660	etc.
3	any calculation that is correct		
4	drawing shapes+symmetral lines		
5	show workings of multiplication		
6	7, 19, 43, 91,187,379	9,25,57,121, 249, 505	11,41,131,401, 1211, 3641
7	children to show examples		
8	show workings of multiplication		

Lesson 5 (page 50)
A: 3 or 9 o'clock; B: triangular prism; C: octagon, 16.

Lessons 5a–5c (pages 51–53)

Q	5a	5b	5c
1	7	8	11
2	2 circles 1 rectangle	2 hexagons 6 rectangles	2 octagons 8 rectangles
3	24	12	20
4	N/A	N/A	N/A
5	N/A	N/A	N/A
6	N/A	N/A	N/A
7	64	192	768
8	any appropriate shapes		

Lesson 6 (page 54)
A: class discussion; B: discussion about statement; C: 225, 449 (double and subtract 1)

Lessons 6a–6c (pages 55–57)

Q	6a	6b	6c
1	87, 88	137, 138	237, 238
2	show workings out for multiplication		
3	$60 \times n$	$60 \times 60 \times n$	$60 \times 60 \times 24 \times n$
4	false	false	false
5	209	156	285
6	examples of Q. times tables		
7	show workings out of addition		
8	show examples		

Problems involving 'real life', money or measures

Lesson 1 (page 58)
A: 500; B: £22.39; C: 8 litres

Lessons 1a–1c (pages 59–61)

Q	1a	1b	1c
1 tapes posters	CD £2,50 £1.08 85p	CD £1.25 £2.08 £1.75	CD £7.50 £4.41 £1.85
2 cream ice crm	180g choc 600ml 6 scoops	225g choc 750ml 9 scoops	420g choc 1050ml 12 scoops
3	£2.28	£4.06	£7.86
4	52	90	170
5	20	29	43
6	41	45	57
7 Highest Lowest	£1.05 57p	£1.20 75p	£1.79 £1.16
8	3hrs, 30mins.	6hrs, 15mins.	7hrs,55mins.

Lesson 2 (page 62)
A: 2kg; B: 184 pages;
C: 4 minutes, 29.3 seconds

Lessons 2a–2c (pages 63–65)

Q	2a	2b	2c
1	£10.40	£18.72	£29.12
2	40 days	67 days	115 days
3	20	58	78
4	114cm	128cm	170cm
5	118	218	338
6	£4.30	£7.27	£18.27
7	£2.30	£2.35	£2.47
8	1250ml	2500ml	4375ml

Lesson 3 (page 66)
A: 77; B: 36 sausage rolls, 54 sandwiches, 24 cakes; C: £16 ($) and £38 (euros)

Lessons 3a–3c (pages 67–69)

Q	3a	3b	3c
1	130 miles	190 miles	240 miles
2	28	18	46
3	£14.70	£17.50	£30.10
4	94	174	254
5	4	11	20
6	1hr, 52mins	2hrs, 4mins	2hrs,38mins
7	35	57	21
8	102	162	242

Lesson 4 (page 70)
A: 16:57; B: Tamsin 36kg, Felix 54kg; C: 2.50m

Lessons 4a–4c (pages 71–73)

Q 4a 1	104 Egyptian pounds,
	986 Jamaican dollars
	133 Hong Kong dollars
Q 4b 1	208 Egyptian pounds
	1972 Jamaican dollars
	266 Hong Kong dollars
Q 4c 1	364 Egyptian pounds
	3451 Jamaican dollars
	465.5 Hong Kong dollars

Q	4a	4b	4c
2	4.4 litres	11 litres	16.5 litres
3	14hrs 5 mins	13hrs 35mins	13hrs 55mins
4	30	49	98
5	7.5km	16km	22.5km
6	4 raffle 8 coconut	4 raffle 8 coconut	8 raffle 16 coconut
7	2340	4420	5460
8	£19.50, £175.50	£39 £156	£58.50 £136.50

Lesson 5 (page 74)
A: 76; B: £9.50; C: £24.90

Lessons 5a–5c (pages 75–77)

Q	5a	5b	5c
1	6m	20m	24.5m
2	£9.95	£15.45	£20.45
3	2.05kg	3.2kg	6.2kg
4	99cm	88cm	71.5cm
5	8.41, no	8.47, no	8.47, no
6	£25.75	£54.75	£67.70
7	38	48	46
8	20	30	56

Lesson 6 (page 78)
A: £67.20; B: run 19.5, walk 6.5 miles; C: 30

Lessons 6a–6c (pages 79–81)

Q	6a	6b	6c
1	$150m^2$	$600m^2$	$832m^2$
2	£14.40	£19.11	£41.40
3	$24m^2$ paved 4 x 2m etc.	$42m^2$ paved 7 x 2m etc.	$64.5m^2$ paved 2.15x10m etc.
4	30	45	65
5	12	28	54
6	156	195	325
7 Change	£11.10 £8.90	£25.65 £4.35	£39.15 £10.85
8	384; 216 to collect	684; 316 to collect	1284; 216 to collect

www.brilliantpublications.co.uk

Printed in the United Kingdom
by Lightning Source UK Ltd.
111185UKS00001B/245-252